THE BEST OF THE AGFA WILDLIFE AWARDS

TWENTY YEARS OF WINNING PHOTOGRAPHY

AWARD-WINNING IMAGES OF AFRICA

TWENTY YEARS OF WINNING PHOTOGRAPHY

Published in 2002 by
New Holland Publishers (UK) Ltd
London • Cape Town • Sydney • Auckland

Garfield House, 86-88 Edgware Road, London W2 2EA
80 McKenzie Street, Cape Town 8001, South Africa
14 Aquatic Drive, Frenchs Forest, NSW 2086, Australia
218 Lake Road, Northcote, Auckland, New Zealand

First published 2000 by
Struik Publishers (Pty) Ltd

1 3 5 7 9 10 8 6 4 2

ISBN 1 84330 360 4

Publishing manager: Pippa Parker
Editor: Simon Pooley
Designer: Dominic Robson
Proofreader: Thea Grobbelaar
Agfa Coordinator: Barbara Garner

Reproduction by Hirt and Carter Cape (Pty) Ltd
Printed and bound by Craft Print (Pte) Ltd, Singapore

Publisher's Acknowledgement: The Publishers would like to
thank Barbara Garner at Agfa for her inexhaustible energy
and enthusiasm for this project and for the Agfa Awards.
Barbara's contribution in tracking down the chosen
photographers was essential in compiling this book.

ENDPAGES: Dandelion Precision **Herman van den Berg** 1995
HALF TITLE: Sunset with Grass **Peter Pickford** 1999
FULL TITLE: Stampede **Johann Knobel** 1992
THIS PAGE: Dew Drop Lenses **JL du Plessis** 1999

Contents

Overall Winner 2000
Lioness Stalking
Gerald Hinde

Agfa Foreword

The majesty of an African lion as his roar reverberates through the undergrowth, truly king of this unique realm he roams unhindered. The earth's rumble as a herd of elephant approaches the water hole, the calves skittering among those giant sheltering legs. A fish eagle's evocative cry high above, as it swoops down to the glistening water below, talons spread for the catch. Two giraffes silhouetted momentarily against a blood-red African sunset. The elusive patterned symmetry of that most beautiful of cats, the leopard, as he stalks his prey by silvery moonlight. This is the ever-changing, breathtaking kaleidoscope of Africa. These are the images seen through the lens of the wildlife photographer, as he too patrols the African bush. Stalking his prey, not with a gun or poacher's snare, but with camera and film, he waits patiently for that one heart-stopping moment when all the elements come together in perfect synergy – subject, light, composition, action – to be indelibly etched in the emulsion inside his camera.

We in Africa have been entrusted with a unique legacy, ours to protect and preserve for future generations. Millions of people all over the world will never be fortunate enough to experience the wonder of the African wild at first hand, and we need to reach out to them, to inform, stimulate and delight the eyes of a wider and less privileged audience. Through the magic of photography, the wildlife photographer, with his talent, skill and dedication, is able to capture on film the unique images of this brooding, mysterious continent we call home, and to share his moving experiences with all of humanity. We hope this book will go some way towards realising this goal.

AGFA ◆ *Agfa*

Portfolio of Winners
1981–2000

Since its inception in 1981, the Agfa Wildlife and Environment Awards has become Africa's most prestigious wildlife photographic competition, attracting entries from top wildlife photographers throughout the world. Entries are invited in eight different categories, including a category for young wildlife photographers of 21 years and under. The Top 50 slides are selected by an independent panel of judges, with 1st, 2nd and 3rd prizes of cash and film awarded in each category. The overall winner of the Agfa Wildlife and Environment Awards is chosen from the winners of each category. Only slides taken in Africa, or the islands or waters surrounding Africa, are eligible, in order to focus on the wealth of wildlife and scenic splendour of this continent, and keep the competition uniquely African in content.

Wildlife photography has gone through many changes in the course of the 20 years of the competition's existence, and in turn, so have judging standards and criteria. Advances in photographic equipment – new technology in camera bodies, lenses and film – have allowed photographers to become more daring. Autofocus, strong portable flashes and high-speed film, as well as 35mm SLR cameras, capable of shooting up to 1/5000 of a second, have helped push the boundaries of quality, enabling photographers to obtain more impressive results.

Judges currently look for work that shows empathy with new styles and standards. Movement is important nowadays, so a static shot is less impressive to judges. A classic portrait will always be appreciated, but the emphasis has shifted towards animal behaviour, and showing the animal in its environment (in line with ecological trends). There is also less of an emphasis on stylistic rules, and more on artistry – how photographers interpret a scene, then use their camera to say something individual. Indeed, the horizons of wildlife photography are continually expanding to accommodate photographers' talents and ingenuity, and it is encouraging to know that the judges of wildlife photographic competitions are among the first to appreciate this.

An exhibition of each year's Top 50 award-winning photographs travels around South Africa to all major centres. A second set is on permanent display in the Kruger National Park.

In preparing this book we had the pleasurable dilemma of selecting a representative collection of only 176 photographs from the 1 000 stunning images – representing 20 years of prize-winning photography – available to us. The book begins with a gallery of the overall winners for each year of the competition's existence, before moving to a more subjective collection chosen from the Top 50 pictures from each year. Each of the images in the book is accompanied by the photographer's name, the title he or she gave to their winning picture, and the year in which the photograph was entered in the competition. Each photograph in the body of the book has been captioned by its photographer, and where possible details of photographic equipment and settings are included, together with the photographers' contact details, in an appendix at the end of the book.

Overall Winner 1981
After the Mating
Edward Lightbody

Overall Winner 1982
Locked Horns
Gus Mills

Overall Winner 1983
Desert King
Richard Goss

Overall Winner 1984
Survivor
WD Haacke

Overall Winner 1985
Cheetah Charge
Koos Delport

Overall Winner 1986
Forced Landing
Terry Carew

Overall Winner 1987
Silent Retreat
Peter Franklin

Overall Winner 1988
The King
Lex Hes

Overall Winner 1989
Skirmish
Daphne Carew

Overall Winner 1990
Owl Landing
JJ Brooks

Overall Winner 1991
Gemsbok and Calves
Rob Nunnington

Overall Winner 1992
Pelican Take-Off
Alan Wilson

Overall Winner 1993
Rush-Hour Traffic
John Brazendale

Overall Winner 1994
Hippo Feud
RJ van Vuuren

Overall Winner 1995
High-Speed Chase
Maria Jacovides

Overall Winner 1996
Crocodile Attack
Nick Greaves

Overall Winner 1997
Lion Charge
Beverly Pickford

Overall Winner 1998
Black-backed Jackal
Kim Wolhuter

Overall Winner 1999
Sooty Tern Colony
Roger de la Harpe

Out of Control **Beverly Joubert** 1995

It was mid-November, and we were sitting watching the lions we work with in Savuti, in Botswana. Because of the cool weather, this cub started to explore up a tree and soon got into trouble! She wobbled and finally lost control altogether. I was lucky to get this because I started laughing and couldn't keep steady.

Mtunzini Frog **Gert Lamprecht** 1995

This specimen of the Forest Tree Frog (*Leptopelis natalensis*) was only around 30 mm from snout to vent. These frogs breed in riverine bush and swamps along the northeastern coast of South Africa. A custom-built twin flash system mounted onto the camera made it possible to photograph this tiny, colourful creature.

Leopard Yawning **M & C Denis-Huot** 1999

This picture was taken in Samburu Reserve in Kenya in 1996. This female yawned and stretched herself after waking up. The big difference in light between her and the bushes resulted in the dark green bushes coming out almost yellow.

Annoyed Elephant **Johan Jooste** 1992

I was sitting in a hide in the Etosha National Park in Namibia, when a youngish gemsbok slipped and fell into a water trough in front of me. He was struggling to get out when two big bull elephants came to quench their thirst. When this elephant saw the gemsbok in 'his drinking water', he started to kick sand toward the unfortunate gemsbok, who was still in the water.

16

African Buffalo **Frank Krahmer** 1999

This photograph was taken in the Aberdares National Park in Kenya. After a great bath in a water hole filled with orange Aberdares soil, this buffalo enjoyed the evening drying in the late sunlight. I have never seen a creature this dirty from top to bottom since.

Interleaved Dead Trees Hendrik Ferreira 1999

During the last few minutes of sunshine, the Dead Pan at Sossusvlei
in Namibia assumes the atmosphere of a sacred place, and visual
poetry develops. For this picture, I waited until the sun touched only
the furthest tree.

Namib Kuns GPL du Plessis 1982

I found this tree near Sossusvlei in the Namib Desert. You need the
early-morning light on the dunes to see the ripples in the sand so
clearly captured in my photograph.

21

Spur-winged Surfer **Albert Froneman** 1997

I spent several days photographing waterfowl at a small farm pond in the southwestern Cape, in South Africa. Numerous Spur-winged Geese were resident in the area and during the midday heat they often retreated to the shallow water to bathe, preen and rest. Eventually this youngster started bathing right in front of me.

Giant Kingfisher Catching a Fish **Beverly Joubert** 1985

At the Linyanti River we were visited regularly by a Giant Kingfisher. He had an amazing diet: one day a striped snake, another day a huge frog. Mostly, however, he caught fish – like normal kingfishers.

Egyptian Goose **William Barr** 1998

The Lower Sabie area of the Kruger National Park must be one of the best places in the natural world to photograph water birds. This Egyptian Goose was feeding on algae growing on the rocks at the dam wall across the Sabie River. The late afternoon light emphasised just how colourful this 'common' bird is.

Heron Bathing **Johan Beyers** 1996

I had been watching this Grey Heron in familiar frozen position, waiting for a fish, for quite a while – when suddenly it apparently decided that the fishing was no good and opted for a bath instead. I managed to grab this shot just before the heron flew off.

Moon between Rocks, Spitzkoppe
Theo Allofs 2000 *(opposite)*

My wife Sabine and I were camped near the Spitzkoppe Inselberg in Namibia when we noticed an elongated crack in the rock on a rugged slope. The moon was already up in late afternoon and it struck me that this formidable graphic structure was perfect to frame the moon. I raced up the slope to a spot from where the moon was positioned in the middle of the crevice, and was lucky to capture a very special moment.

Spitzkoppe Arch **Thys van der Merwe** 2000

I spent two days camping at Spitzkoppe in Namibia, attempting to capture the magical light on these awe-inspiring rock formations. On the morning of the second day, the clouds parted for a short while, allowing me to capture a brief moment of magic at the Arch.

Lion Carrying Cub **M & C Denis-Huot** 1999

This was taken in the Masai Mara in Kenya, in August 1998. Two females of the pride we were following had young cubs. One morning, they began to carry the eight cubs from one hide to another, about one kilometre distant. The light was beautiful.

River Croc and Hatchling **Pat de la Harpe** 1996

After the young crocodiles hatch, the female Nile Crocodile carries them down to the water and relative safety. The adults can be fairly tetchy during this time!

Adult Hyaena with Pup **Anne & Steve Toon** 1998

Early one morning at a Spotted Hyaena den in the Kruger National Park we saw this female, newly returned from a night's foraging, playing with her cub. We wanted to capture something of the gentler side of the hyaena's nature, belying its popular image as a bloodthirsty scavenger. Spotted Hyaenas make excellent mothers, with a very high success rate in rearing their young.

Springbok in Rain **Heinrich van den Berg** 1997

I was fortunate to experience a sudden downpour in the Auob River in the Kalahari Gemsbok National Park (joined to Botswana's Gemsbok National Park to form the Kgalagadi Transfrontier Park in May 2000). A herd of springbok turned their backs towards the rain while opening their crests slightly. The light came through while it was still raining and I took this picture against the light.

This Little Piggy... **C Anniciello** 1993

Early one overcast morning we came across a mother leopard teaching a young leopard the art of eating breakfast from a 'warthog take-away'. Three baby warthog were pulled out of the den one at a time, and after playing with them, mother ate one and son the balance.

Surprise Encounter **Beverly Joubert** 1995

A few days after heavy summer rains these huge bullfrogs wriggle their way out of the earth and start kicking up a racket. Jackals of course investigate every little rustle in the grass. This one didn't expect this quite ferocious frog to emerge!

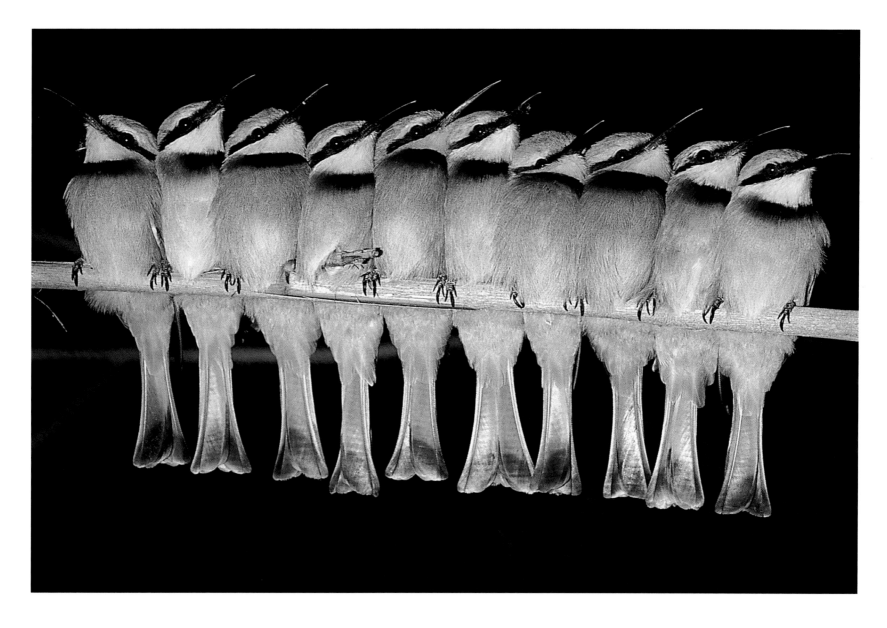

Previous pages

Coral Trout **Graham Wallington** 1993 *(top left)*

This photograph was taken in Sodwana Bay, one of southern Africa's premier diving destinations, on the north coast of KwaZulu-Natal.

Ambush **Colin Mostert** 1998 *(bottom left)*

There are many species of extremely local, well-camouflaged kingfish in the cold Cape waters. This species, the Chinese Klipfish, is superbly adapted to blend into its preferred micro-environment of Crinoid Starfish (feather stars).

Big Eyes **Colin Mostert** 1998 *(top right)*

The common West Coast Klipfish can be found in many interesting colour forms. This animal's beautiful eyes can only really be appreciated with the use of macrophotography and an artificial light source.

Coral Hind **Reimund van Eeden** 1997 *(bottom right)*

I was on a dive and took up a position hovering in front of a cleaner station. This is an area where fish go to be cleaned by Cleaner Shrimps and Cleaner Wrasses. I discovered this Coral Hind being cleaned by a shrimp and captured him on film at a depth of around 25 metres.

Ten Little Bee-eaters in a Row
Terry Carew 1994 *(opposite)*

While poling through a hippo run in the Okavango Delta one evening, we came across these Little Bee-eaters on a papyrus reed. They were huddled together for warmth and had weighed down the reed until it was horizontal. I used a long lens to avoid frightening them.

Flycatchers in the Cold **Tim Jackson** 1994 *(above)*

One chilly morning in the Kalahari I found this group of Marico Flycatchers tightly huddled up for warmth in order to survive the overnight temperatures, which had dropped to below freezing. They were enjoying basking in the first warm light of the morning when I took this opportunity to photograph them.

Rage **Richard du Toit** 1999 *(opposite)*

I had spent nearly an hour following a leopard at Mala Mala when she encountered another female in a thicket. Almost immediately they clashed and tumbled, fighting furiously for a few seconds. I was in a perfect position to shoot as a shaft of late-afternoon light painted the battling cats a liquid gold.

Jackal Feud **Helmut Niebuhr** 1996 *(above)*

This impressive male jackal (right) dominated a springbok carcass, allowing only a female and two juveniles to feed. Intruders were aggressively chased away. This scene was shot in the Kalahari Gemsbok National Park.

Fatal Attraction **Pat Donaldson** 1992 *(right)*

Following the death of a dominant male lion, all the pride's cubs are killed and the new dominant male must mate quickly to establish his own genes within the pride. The female is particularly aggressive at this stage and the mating attempt shown here was unsuccessful. She did, however, eventually succumb to his charms!

Cheetah **Helmut Niebuhr** 1995

Following his mother's example, this sub-adult joined in stalking springbok in the Kalahari. It came as a pleasant surprise to see how effectively the essence of what was a very intense moment was captured in this photograph.

Hokaai! **Cornie Malan** 1996

It was a rainy and overcast day near Skukuza in the Kruger National Park, when a startled baboon family ran past our vehicle, and I was fortunate in getting this panned shot.

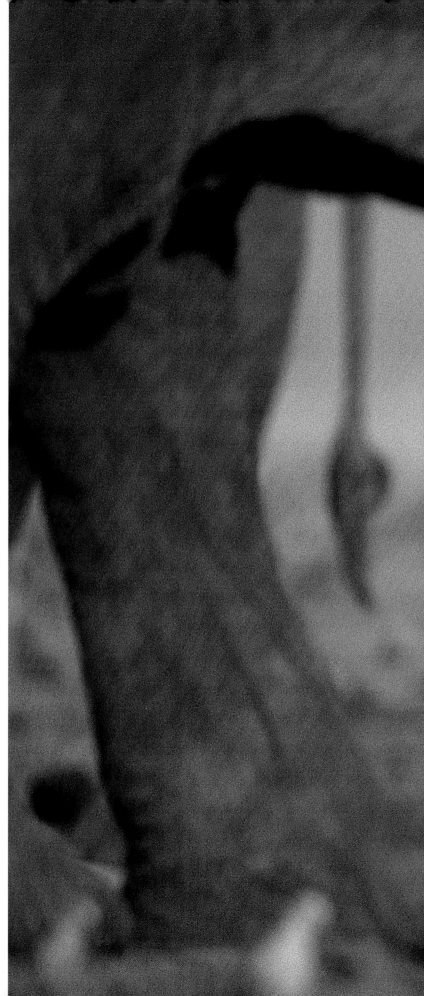

Last Light Drink **Wayne Matthews** 2000

This photograph was taken at Mahlasela Pan in Tembe Elephant Park on the border of KwaZulu-Natal and Mozambique. I was sitting in a hide at the edge of the pan when this lone bull came down to drink, just as the sun was setting.

Black-backed Jackal behind Elephant Leg
Theo Allofs 1998

This picture was taken in the dry season at a water hole in the Savuti area of Botswana's Chobe National Park. It was early morning, shortly after sunrise, the time when the thirsty animals were coming to drink. A number of elephant were already gathered at the water hole. Huge flocks of doves were arriving all the time, to drink unconcernedly and busily among the elephants' feet. The jackal was intent on killing doves – which we had observed him do before. He was hiding behind one elephant's leg, intently watching the doves and waiting for the right moment before darting out and trying to snatch one.

White-winged Terns, Uganda
Patrick Wagner 1997

Patrick photographed these graceful birds in June 1996, on the Kazinga Channel in the Queen Elizabeth National Park. Patrick was experimenting with slow shutter speeds at the time.

Pink-backed Pelican Duo **DC Williams** 1997

I had been trudging wearily through the thick mud near the Umgeni River mouth in Durban in KwaZulu-Natal, trying to get closer to the pelicans. The tide had turned and I had just a short time left to get the shot. Miraculously, these pelicans took off in perfect unison.

Black Wildebeest and Newly Born Calf
Herman van den Berg 1994 *(above)*

As we approached a herd of Black Wildebeest at Midmar Game Park, the whole herd scattered except for one wide-eyed cow. Then we noticed the newborn calf lying in the grass. A short while later the calf stood up and was able to run away with the mother.

Red Lechwe and Young
Terry Carew 1994 *(opposite top)*

This young Red Lechwe was feeding from its mother with great enthusiasm, at times lunging so forcefully that her hind legs were lifted off the ground. This picture was taken in Moremi in Botswana.

Mother Care **Piet Heymans** 1993 *(right)*

In 1993 the rains came to the Kalahari in mid-December and January. Springbok were plentiful and the bulk of ewes gave birth within a two-week period. A number of lambs were in a crèche with nurse-maids, and I got this photograph at feeding time.

Black Rhino **Uwe Anders** 1993

I took this picture on my third day in Africa, while on a foot patrol with Parks Board rangers. We got very close to the rhino and only later did I realise that the situation was very critical. The Zulu name of this female rhino is *Maphosa*, which means 'the one who chases people'.

Elephant and Gemsbok **Jill Sneesby** 1998

I used a low viewpoint and cropped composition to emphasise the size of the elephant. The photograph was taken in Namibia's Etosha National Park.

Antics Nico Myburgh 1994

The Plum-coloured Starling's colour changes as the sun climbs higher, so at every visit you get a different colour rendering. The correct colour only shows for a short time at about 09:00 and again at 16:00. For this picture I set up two flash heads to recreate this lighting.

Common Agama Bernard Castelein 1999

It was only after I had spent half an hour trying to take a picture of another, much shyer, Common Agama that I noticed this male sitting less than two feet above my head.

Red Hartebeest in Flight
Herman van den Berg 1994 *(left)*

One of the roads at Midmar Game Park in Kwazulu-Natal runs parallel and quite close to the Midmar Dam's edge. This provides excellent opportunities for photographing animals on the run because they don't run away from the vehicle, but parallel to the vehicle.

Airborne Impala **Koos Delport** 1981 *(above)*

I was photographing a herd of impala along the edge of the Allemanskraal Dam, near Winburg in South Africa's Free State province, when this one suddenly fled. My photograph captured her in spectacular mid-leap.

Pronking Springbok **Barrie Wilkins** 1984

It was late afternoon in the Kalahari when I captured this springbok lamb doing its presunset pronking. It is not clear why springbok perform these exuberant, often vertical leaps into the air – one theory has it that it is to let predators know that the animal is fit and fast.

Two Lion Brothers **Gabriela Staebler** 1996

The lion brothers were sleeping in the middle of the plains of the Serengeti, not far from Simbakopjes. I watched them from some distance. Suddenly a herd of zebra passed, and the lions woke up and were completely alert and interested.

Following pages

Cheetah Cub Lookout
Luke Hunter 1995 *(left)*

I spent four years studying cheetahs in KwaZulu-Natal where they had been re-introduced after a 50-year absence, and these cubs were the first born in the new population. Cheetahs are poor climbers but trees often feature in the cubs' frenetic games. These three raced frantically up this tree and paused just long enough for me to take this shot.

Leopard on Termite Mound
Martin Harvey 1999 *(right)*

I was following this young male leopard in a private game reserve. It was obviously in a playful mood when it suddenly jumped up onto this termite mound. It was only there for seconds, as I frantically shot off a few pictures.

Webfooted Gecko *(opposite top)*
Roger de la Harpe 1999

These little geckos live in the sand dunes around Swakopmund in Namibia, and we were lucky enough to find one early in the morning when it was still cool and the gecko fairly sluggish.

Watch Out! **RJ van Vuuren** 1997 *(opposite bottom)*

This slide clearly depicts the behaviour of the Horned Adder, showing it burying itself in the sand. Capturing the protruding tongue in sharp focus gives plenty of impact, and this snake's excellent camouflage is also shown.

A Tree in the Desert **Henk Loots** 1995

We had high hopes of seeing the sun rise over Sossusvlei in the Namib Desert, but did not leave the campsite at Sesriem Canyon early enough. An unexpected bonus, however, was coming across this interplay of light and shadow on a perfectly shaped dune and a gnarled thorn tree a few kilometres before we reached the vlei.

Namib Desert **Marike Bruwer** 1997

I visited the Namib Desert in 1997, after the rains in Sossusvlei, to photograph this rare sight. The clouds added something special to the landscape of sand that was revealed to me that afternoon when I reached the top of the high sand dune in the vlei.

Shadows on Namib Dunes **Matthias Graben** 1998

I was on my way to Sossusvlei late one afternoon, and the sun had already created dark shadows on the dunes. The huge dune in the background was like a dark curtain behind the smaller ones in front, and the trees reminded me of toys standing in the scenery.

Cheetah Mother with Cub **Gabriela Staebler** 1996

As dusk fell in the Masai Mara, this cheetah left her three cubs under a tree and climbed a large termite mound for a better view of potential prey. For a few seconds the sun broke through the clouds, just as one of the cubs turned up and sat down beside her.

Suricate with Young **Peter Lillie** 1992

I photographed these suricates in summer in the Kalahari Gemsbok National Park. These mongooses live in groups and favour arid environments. They spend the daylight hours digging for reptiles and invertebrates, and take turns watching over the youngsters.

Sad, Wet and Sorry **Jamie Thom** 1994

Rain is seldom good for photography, but after rain can be amazing. I was driving in South Africa's Kruger National Park when I saw this troop of baboons, looking sad, wet and sorry. I leaned towards the window and snapped two shots – the baboons' expressions definitely make the shot for me!

A Leisurely Lookout **Helmut Niebuhr** 1996

After entertaining us with its playful antics, this female leopard climbed to a vantage point and settled, gazing unperturbed at the camera. I photographed her at Biyamiti, in the Kruger National Park.

Ousus Piet Heymans 1994

I have been following this lioness since she was a cub, and first photographed her in January 1993, in pouring rain at Kousant in the Kalahari Gemsbok National Park. This photograph was taken at Bedinkt, in a green Kalahari during January 1994 when she was one year old. She gave birth to cubs in February 2000.

Parson's Chameleon Eye **Martin Harvey** 1997

This is the world's largest chameleon, but I was more intrigued by the amazing colours and patterns on its skin than by its size.

Flower Power **Colin Urquhart** 1995

This was taken in early spring 1994, in the Addo Elephant National Park near Port Elizabeth in South Africa. The display of daisies was magnificent that spring, and makes for a very unusual picture. Addo's elephants are in general friendly, and very accessible to tourists, which means it is often not necessary to use long lenses to photograph them.

African Honey Bee **Jill Sneesby** 1995

This photograph was taken in a sunflower field in the Free State. I went in to photograph the sunflower, but was attracted by the colour combination as the bees flew around.

Fact of Life **Helene Heldring** 1995 *(opposite)*

We had followed a single pack of wild dogs in Botswana's Okavango Delta for months and they became habituated to our presence. One morning we managed to keep up with them during their hunt and were on the scene when they took down this male impala.

Aggression **Wayne Griffiths** 1995 *(above)*

I had spent five days at the carcass of a giraffe that had been pulled down by lions in the Kruger National Park. When the lions moved off, the hyaenas took over, jealously guarding the remains of the carcass from the attentions of the waiting vultures.

A Bloody Mess **Gus Mills** 1983 *(right)*

I had followed these two Spotted Hyaenas for more than six hours and 20 kilometres before they pulled down this yearling wildebeest. The fast, 50 kilometre-per-hour chase lasted 1.2 kilometres.

Namaqualand Wild Flowers **Theo Allofs** 1998

While on a journey through South Africa's Northern Cape province, I considered myself lucky to experience the fascinating spectacle of the wild-flower bloom in Namaqualand. Amid the explosion of colour all around me, this unusual natural pattern of a single bright orange flower in a carpet of velvety purple caught my eye.

Nosestripe Clownfish, Mozambique
Patrick Wagner 1997

This was taken at Benguerra Island off Mozambique. Patrick sat with the frame over the anemone, while the clownfish danced in and out. The surge was moving the anemone around when it suddenly parted to reveal the fish sitting underneath and he snapped the shot.

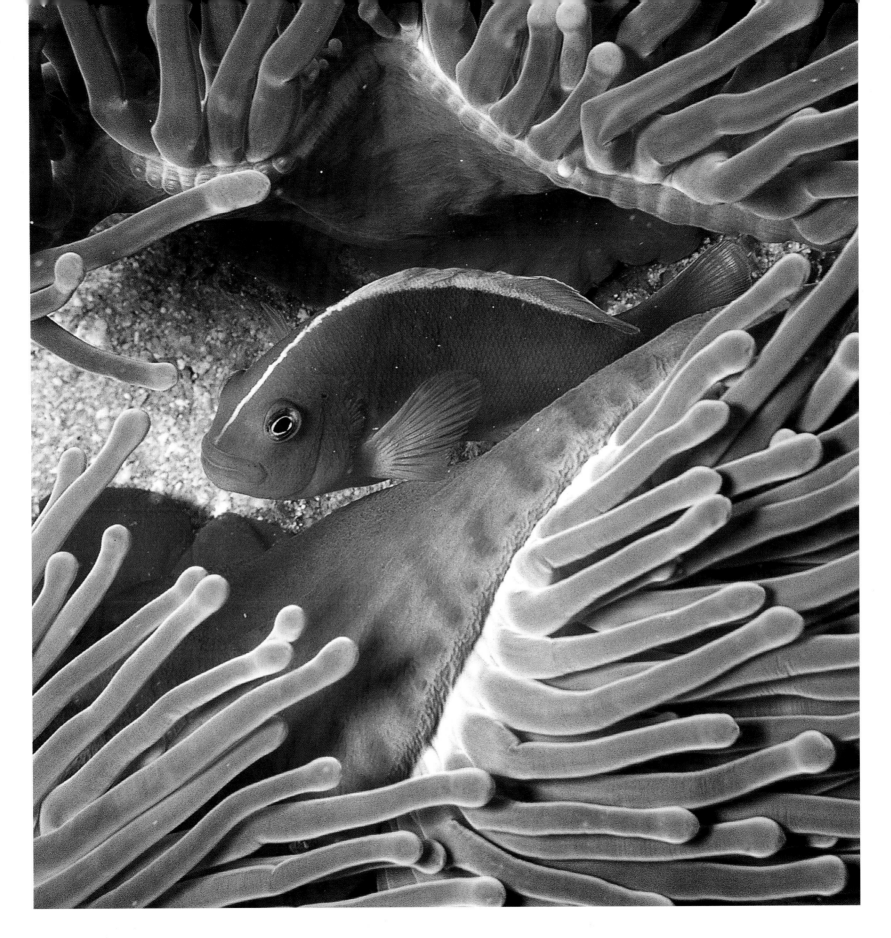

Quelea Dance **Beverly Joubert** 1994

When the quelea mass together in these huge flocks it is one of the more exciting events in the bush. But you have to be quick to get good images. I spot-metered the highlights and exposed for that, letting the shadow areas just do what they do.

Lioness Wading **Jill Sneesby** 1997

This photograph was taken in Etosha National Park during the wet season. The lioness wanted to cross over and had no option but to go through the water. You can see how very wary she was.

74

Red Lechwe Crossing Flood Plain
Beverly Pickford 1997

To fly over the Okavango Delta in a microlight is an incredible experience that I found difficult to translate onto film. These Red Lechwe, caught in the first light of the winter morning, perhaps express it best. My greatest difficulty was to get my fingers to function in the extreme cold!

Eekhoring **Johan Louw** 1996

I captured this Tree Squirrel at the Kanniedood Dam near Shingwedzi in the Kruger National Park. Two of them were sitting on a tree in the late afternoon sunshine. Nature's camouflage is great, as can be seen in the warm tones of the afternoon sun against the bark of the tree and the colour of the squirrel's skin and hair.

Elephant Shrew **Richard du Toit** 1997

I found this Elephant Shrew one winter morning in the Auob River bed in the Kalahari. He spent a few nervous moments in the open, twitching and grooming, before disappearing. Amazingly I found him sunning himself on the same log on my next trip three months later!

Johnstone's Chameleon **Martin Harvey** 1999

While photographing in Central Africa I found this chameleon in a bush right outside my hotel room. It was totally unconcerned by my presence and let me get right up close to photograph it.

Kermit was Here **Marius Burger** 1997

'Kermit' was photographed in the rainforest of Perinet in Madagascar, one of the world's frog diversity hotspots. Nocturnal ramblings in pursuit of these delightful creatures have provided some of the most enjoyable moments of my life. Capturing Kermit's image on film presented no major photographic challenge – it was simply aim, focus and shoot, with Kermit's good looks doing the rest.

Robber Fly **Terry Carew** 1985

This is a powerful and aggressive predator which often catches its prey on the wing. As can be seen from the stout proboscis and enlarged thorax housing the wing muscles, the Robber Fly is a force to be reckoned with in the insect world.

Hamerkop Scratching **Joan Ryder** 1985

I spotted this Hamerkop on a dead tree overhanging the water at
a dam in the southern region of the Kruger National Park. Because
I approached quietly and cautiously, the Hamerkop remained in
position, preening and scratching.

Dog Kill **Peter Pickford** 1997

Wild dogs are among my favourite predators, but they are very difficult to photograph in action because of the vast distances they cover during the chase. Two months of persistence paid off when these dogs chased this lechwe full circle to catch it within 100 metres of our vehicle.

Pink-backed Pelican **Herman van den Berg** 1994

As pelicans are very skittish, I approached this magnificent bird – perched on a tree stump in the Umgeni River estuary in Durban – in a floating hide, until I was well positioned to photograph it. Photographing birds in this way is very exciting and rewarding.

Albatross Chick **Jaco van Wyk** 1995 *(opposite left)*

I spent 13 months on subantarctic Marion Island doing research for the Department of Environmental Affairs and Tourism during 1993–1994. While working one late winter's day, the sky suddenly cleared and the setting was perfect for this photograph.

Preening Spoonbill **Philip van den Berg** 1993 *(opposite right)*

I was photographing weavers at the pool in the Durban Botanical Gardens, when a spoonbill landed and started preening itself. I used a flash to fill in the areas not lit by the early morning sunlight.

Following pages

Wet Lion **Freek van Eeden** 1998

I saw this wet lion and his friend at a water hole south of Nossob Camp in the Kalahari Gemsbok National Park. The lions looked miserable and tired and took turns to cool themselves underneath the overflow of a fibreglass dam.

Wildebeest **Adrian Bailey** 1998

It is generally quite difficult to get really close to wildebeest. While working on a book project in Moremi Game Reserve in Botswana several years ago, I came across this bull who seemed to stand at the same spot, in a spectacular spot of light next to the track, every evening. It was difficult not to get a great portrait.

Sandstorm Fantasy **Len Miller** 1992

On a trip to Sossusvlei a sandstorm made its appearance late one afternoon. It transformed the landscape into a fairyland of wonderful mood. I was most fortunate to find a dead tree of great aesthetic appeal as the centre of interest for my photograph.

Homeward Bound **Jill Sneesby** 1985

This photograph was taken in the Kalahari Gemsbok National Park and, as can be seen from the amount of dust around, it was taken during a very dry period when the Kalahari was just a dust bowl. Thousands of springbok came down into the valley to find water, and this shot shows them on their way to the water hole.

Zebras in a Panic in a Pan
Kim Wolhuter 1999

At dawn on most days the zebra came to drink at this pan in the Etosha National Park. Enjoying the water, they always had to walk in at least knee-deep, but if there was any slight suggestion of danger the whole herd scattered for dry ground.

Following pages

Hunter and the Hunted
Geoff Spiby 1999 *(left)*

This photo of the Red-mouthed Grouper (hunter) and the Glassy Sweepers (hunted) was taken at Brothers Islands in the Red Sea. My strobe had malfunctioned and I was forced to look for silhouettes at that depth (around 15 m). Divers moving under an overhang flushed these fish out into the open.

Plunging into the Deep Blue
Stefania Lamberti 1999 *(centre)*

I spent a few years filming in the Indian Ocean for several wildlife programmes. The turtles of Maputaland were the subject of one of our documentaries. This one followed us on our ascent and, while we were at our safety stop, she went to the surface, took a breath of air, and swam down again past us.

Goldies **Geoff Spiby** 1999 *(right)*

Scuba divers' bubbles scare goldies into the safety of the reef. I shot this photo on snorkel at Erg Abu Ramada in the Red Sea. I lay quietly next to the reef, and the goldies slowly moved further from the reef, eventually filling my viewfinder. This was shot at a depth of about six metres.

Fighting Giraffe Bulls **Ingrid van den Berg** 1998

I was photographing giraffes early one morning when two young bulls started necking (their way of sparring). The light was perfect and I managed to get quite a number of shots.

Karoo Bigfoot **Trevor de Kock** 1993

While shooting *The Great Karoo – A Secret Africa* for the BBC's Natural History Unit, I had the opportunity to get this unusual close-up of a male ostrich, whose territory we were clearly invading. My assistant kept the bird at bay by holding a tripod box aloft so I could lie on the ground. It was not the thought of the ostrich's dangerous kick that worried me so much as what might fall from above!

African Wildcat **Koos Delport** 1989

The African Wildcat is a shy creature which is very rarely seen by day. I spent many hours photographing these small, solitary cats in the Kalahari Gemsbok National Park.

Long-crested Eagle **Marco Calore** 1998

This portrait, showing a Long-crested Eagle calling its mate, was photographed in Mpumalanga province, South Africa. These striking raptors are fairly common in the eastern parts of the country.

Lion Yawning **Nigel J Dennis** 1998

This magnificent lion lives in the Sabi Sabi Private Game Reserve, adjacent to the Kruger National Park. I have photographed him on many occasions – in fact every time I see him it is rather like meeting an old friend. I used a long lens to minimise a distracting background.

Elephant Taking Mud Bath **Theo Allofs** 1998

I encountered this elephant bull frolicking in one of the few remaining mud pools near Savuti, in Chobe National Park in Botswana, at the height of the dry season. Thousands of Cape Turtle Doves who had come in to drink were swirling overhead, but the elephant and the birds ignored each other. Judging from the expression of the elephant wallowing in the cooling mud, he was experiencing a moment of pure pachyderm bliss.

Leopard with Baboon Kill Martin Harvey 1999

I came across this leopard struggling to wedge its kill into a fork in a tree. As the kill started to fall out, it just managed to catch the carcass with its claws. Leopard drag their kills into trees to keep them safe from other predators such as hyaena and lion.

Feeding Vervet Monkey Philip van den Berg 1997

I was photographing life in the mangroves in the Umlalazi Nature Reserve on the coast of northeastern KwaZulu-Natal, when a troop of vervet monkeys arrived. One hung upside down for a short while, feeding on a seed. I managed to take two shots.

Knysna Lourie **Ray Hill** 1983 *(above left)*

This bird was a regular visitor to our garden at Nahoon Mouth, East London. It was attracted to the garden by the fruit of the indigenous Gwenya Tree and Cape Fig, making its presence known by its raucous call. The Knysna Lourie is a shy bird, rarely flying above the tree tops.

Tawny-flanked Prinia **Helmut Niebuhr** 1992 *(above)*

Soft light and the delicate blade of grass complemented this dainty little bird. The photograph was taken at the Kanniedood Dam, near Shingwedzi in the Kruger National Park.

Malachite Kingfisher with Catch **Philip van den Berg** 1997 *(left)*

I was photographing game from Msinga Hide in Mkuze Game Reserve in KwaZulu-Natal when I noticed this kingfisher catching prey and returning to the same perch to feed. I prepared myself and managed to get this shot.

Masked Weaver **Warwick Tarboton** 1994 *(above)*

Masked Weavers use their incomplete nest-shells to display to females and entice them to use one of their nests. This male, which I photographed from a hide near Nylsvley, was intent on charming one particular female.

Red Bishop Calling **Jack Weinberg** 1985 *(above right)*

This picture was taken early one summer morning at Marievale Bird Sanctuary near Johannesburg, South Africa. The Red Bishop had completed building a nest, and was now calling for females to inspect his work.

Fiscal Shrike Perched on Aloe **Philip van den Berg** 1999 *(right)*

While photographing sunbirds feeding on aloes in the Tsitsikamma National Park, I noticed a Fiscal Shrike perching on the flowers waiting to catch insects attracted by the nectar. It gave me ample opportunity to photograph it to my heart's delight.

After the Floods **Colin Mead** 1997

Sossusvlei in the Namib Desert is always a great place to photograph, but in 1997, after the Zebra and Tsauchab rivers came down in flood following torrential rain in the Naukluft Mountains, it was absolutely incredible. This photo was taken 10 minutes after sunrise.

King of the Dunes **Jack Weinberg** 1987

This picture was taken one cold morning in the Kalahari Gemsbok National Park. From the top of this dune the lion had an excellent view of the river bed, while warming himself at the same time.

Following pages

Colour and Cloud **Gerhard Dreyer** 1994

This photograph was taken at a time when I was concentrating on individual flowers, but on this day the patterns of the clouds caught my attention. The picture was taken near the entrance to the Postberg Nature Reserve on South Africa's West Coast.

Coral Kingdom **Colin Mostert** 1995

This lionfish was inhabiting the superstructure of a shipwreck which forms a fascinating artificial reef. I managed to get the camera underneath the fish, and shot directly up towards the surface.

Gemsbok against Dunes **Jill Sneesby** 1987

The simplicity of the dune and sky is what attracted my attention and it was fortunate that the gemsbok was there, giving scale to the dunes. I photographed this scene in Sossusvlei.

Namaqua Chameleon **Nigel J Dennis** 1998

The Namaqua Chameleon is a terrestrial species living in harsh desert areas – not the kind of place you would expect to find a chameleon! To show both the creature and its rather amazing environment I used a super-wide-angle lens. I had to crawl about on the sand to get the right angle – while trying not to get sand into the camera!

Alert Cheetah **Elanie de Villiers** 1998

This photo was taken in the Kalahari Gemsbok National Park, near Mata-Mata, just past Siteas Water Hole. We had followed the mother and cubs for three days, knowing she had yet to catch something big. Then she saw a steenbok on the other side of the river bed...

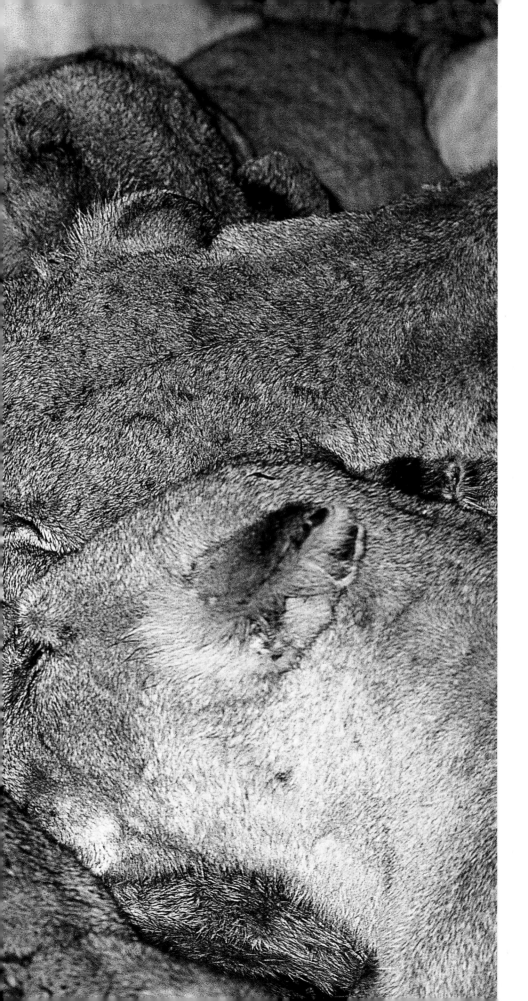

Feeding Frenzy **Jamie Thom** 1994

I was fortunate enough to spend some time with a friend who was studying lions in the Kruger National Park, and this was the first lion kill I witnessed. Sad, excited and scared were only some of the emotions I felt that night as 15 lions ate this wildebeest alive!

Jackal Defending Skeleton **Gabriela Staebler** 1998

One morning in the Masai Mara, I waited to see what would come to feed on the bloody skeleton of a wildebeest. This Black-backed Jackal arrived, but it couldn't feed since it was too busy guarding the food from vultures. After half an hour the jackal gave up, and the vultures took over the kill.

Salticidae in Lair **Les Oates** 1994

On a stroll in the garden I just happened to spot this jumping spider
hiding behind a petal. I rushed to get my camera, but I was too late
to get the jump as the flash had to charge up.

Sea Star Seychelles **Norbert Wu** 2000 *(opposite top)*

This photo shows a Candy-striped Sea Star on Acropora coral. Sea stars (or star fish) are echinoderms, a class of animals that features radial symmetry and includes sea cucumbers and sea urchins. Starfish are voracious predators, feeding upon almost anything that can't move fast enough to escape their grasping tube feet and eviscerated stomach.

Terns **Martin Livingstone** 2000 *(opposite bottom)*

These terns were settling down for the night in the bird sanctuary at Lamberts Bay on South Africa's west coast, and were all facing into the strong southeasterly wind.

New Beginnings **Gareth Hopkins** 2000 *(above)*

In the late winter of 1999, a wild fire devastated our farm in Dullstroom, and while walking afterwards I noticed beautiful flowers growing out of the burnt earth. I realised that life can actually come from death, and got the idea for my photograph New Beginnings. Just then a cloud passed in front of the sun and the colour of the metallic burnt wood changed to a subtle shade of blue-grey.

Koggelmanderwyfie **PC Zietsman** 1995

I photographed this agama preparing her nesting burrow in the Kalahari Gemsbok National Park. When she saw me she got out and came to rest in the shade directly beneath my camera, where I was lying close to the burrow. I picked her up and put her back in the burrow, but she just turned around and hurried back to the only available shade.

Open Wide **Peter Pinnock** 1998

Different fish species come together at 'cleaning stations' to perform acts of mutual benefit: clients have their parasites and damaged skin removed by the hosts, who in return receive a free meal. Here a Goatfish is attended by a Cleaner Wrasse and a Diana's Hogfish.

Blenny in its Coral Home **Malcolm Hey** 1999

During a dive in the Red Sea off Egypt I came across this little Comb-tooth Blenny poking its head out of its resident hole. It constantly bobbed in and out of the hole, so I had to exercise some patience before I was rewarded with the shot that I had in mind.

Springbok Herd **Nigel J Dennis** 1998

I was photographing springbok in the Kalahari Gemsbok National Park, when a cheetah appeared over the crest of a nearby dune. The herd panicked and ran in a tight group. Rather than try to freeze the action, I opted for a slow shutter speed to enhance the feeling of movement.

Cheetah Chasing Springbok **Ivor Migdoll** 1996

I was returning to Mata Mata Camp in the Kalahari late one afternoon when I spotted a cheetah sitting waiting for a springbok which was walking down a river bed. When it neared, the cheetah gave chase. I put in a new roll of film and switched on motordrive, shooting 17 exposures in a few seconds.

Baby Elephant **Jill Sneesby** 1996

I photographed this scene in Addo Elephant Park, in South Africa's Eastern Cape province. I was using a long lens to get in as close as possible, and included the trunks and legs to give a sense of scale.

Keeping Watch (Only Just) **Roger de la Harpe** 1996

We picked up these lions, a number of females and three or four cubs, late in the afternoon, just as they were beginning to wake for the evening's hunt. This little chap could hardly keep his eyes open.

Mother and Foal **Jack Weinberg** 1986

This picture was shot at at 06:30 in Krugersdorp Game Reserve, in early summer. The mother was very protective and, ever watchful and alert, never let the foal stray far from her side.

Migratory Butterfly **Joan Ryder** 1985

Macrophotography is a great challenge, as incredible patience is required. I positioned a powerful flashgun to the left of the butterfly's head to give texture to the scales, and a crumpled piece of tinfoil below the insect to reflect light upwards.

Butterfly Perch Ken Woods 1997

This photograph was taken early on a cold morning when there was no wind, in an area of grass next to a pond in Newlands Forest in Cape Town. The butterfly was sitting in the grass there, and was extremely still because of the cold.

Leopard **Johan Kloppers** 1983

This picture was taken in the Kruger National Park, on the Lower Sabi road. The previous day I had seen a young female leopard drag an impala into a tree, and I suspected she would still be in that area. Sure enough, she returned to her kill, and after she had finished her meal she reclined on the branch looking straight at the camera.

The King **Iky Plakonouris** 1992

We had been waiting for two days to see this lion eat off the carcass of this elephant. The elephant was not hunted by the lion – it had died of natural causes. The shot was taken at around 17:45, in Namibia's Etosha National Park.

Ibis Feeding on Maggots **MJ Livingstone** 1999

The rising waters of Lake Kariba in Zimbabwe had covered the grass of the flood plains, which left little for the buffalo to eat. Too many were dying for the predators to devour them all, and these Sacred Ibis and a Cattle Egret were feasting on a maggot-infested carcass.

The Duel **Seth Hirschowitz** 1996

This photograph of two male buffalo challenging each other was taken one winter's morning at a water hole in the Krugersdorp Game Reserve in South Africa's Gauteng province.

Waterbuck Challenge **Michael J Rooney** 1995

The loud cracking of horns and flying dust on the horizon drew me to this spectacular scene near Lake Kariba in Zimbabwe. These two raging males did not even notice my approach, and their fifteen-minute duel is the most dramatic waterbuck behaviour I have seen.

Springbok Death **Jill Sneesby** 1994

This was taken in the Kalahari, after rain, when there was still a fair amount of dust but the ground cover was starting to grow again. This fight was more than just the usual rutting we see; these males were deadly serious, although the loser did manage to get away.

126

Buffalo Sunset **AJ Rankin** 1996

I was following a herd of about 3 000 buffalo on the Duba Plains in the Okavango Delta. The dust was unbearable but made for stunning photography. It is quite something to be in the middle of a large relaxed herd, amid the noise, the smell and the activity.

Rhinos at Sunset **Gareth Hopkins** 2000

The light was fading fast when we came across a male white rhino trying to mate with a very agitated female. Making matters worse was a large calf who took umbrage to the male's advances on his mother. There was a lot of dust and by positioning myself behind the rhinos and shooting straight into the sun I got the animals silhouetted in the orange dust.

Pounce **Scott Steckbauer** 1992 *(opposite top)*

A Cape Fox pup (*Vulpes chama*) pounces on a lizard that its father brought to it. I observed this pup and its parents for several days while working in Etosha, in mid-December 1989. Their den was just east of Sprokieswoud in Namibia's Etosha National Park.

Lion Cub **Herman van den Berg** 1997 *(opposite bottom)*

We arrived at Rooikop drinking hole in the Kalahari Gemsbok National Park early one morning, and noticed a lioness and two cubs drinking. Afterwards, they fortunately moved in our direction and I managed to take a few shots of this cub that seemed to have fed very well.

Cheetah Cub on Tree **Gabriela Staebler** 1998

A cheetah family was resting under a tree where they had eaten their kill, a Thomson's Gazelle. After sleeping for a while, one of the three cubs climbed onto the big roots next to the tree trunk to have a better view of the herds of zebra and wildebeest in the savanna.

Newborn Baboon **Ingrid van den Berg** 1994

We were watching a family of baboons sunning themselves next to the road close to Skukuza Camp in the Kruger National Park when I noticed a baby baboon being caressed and suckled by its mother. I attempted to capture the contented expression in its face.

Heaven Scent **Martin Kitzen** 1995

Driving through the Etosha National Park, I encountered a jackal strolling beside the road. It stopped and glanced at me through the thicket, then turned and briefly smelt the bush. All this happened in a split second and I just had time to capture this shot.

132

Sunset Vigilance **Franz Jesche** 1997

As the sun was setting over the plains of Deception Valley in the central Kalahari in Botswana, in April 1997, I saw this predatory Black-backed Jackal keeping a late afternoon vigil in the hope of finding and catching a tasty rodent.

Great Egret **Johan Beyers** 1996 *(above)*

This Great White Egret was fishing in a shallow pond, moving around as it followed the fish, and spreading its wings to assist. It is thought that egrets spread their wings to form shadows in which the fish shelter, resulting in easy prey for the bird.

Goliath Heron **Ivor Migdoll** 1993 *(opposite top)*

I was at the Umgeni River mouth early one misty morning when I spotted this heron. I crept closer in my portable hide, and as the mist lifted the bird slowly spread its wings and turned to face the sun, presumably to dry out its wings.

134

Black Egret Shading **Ian Michler** 1998 *(right)*

Black Egrets are fascinating birds to photograph because of their unique 'shading' behaviour. They do this to shut out reflection from the water surface, the better to see fish. This photo was taken from a hide I built on the Nxamaseri Channel in the Okavango Delta.

St Lucia Reedbuck on Sandspit **John Dalton** 1993

I took this photograph late one evening in Catalina Bay, on Lake St Lucia. There was a drought on, and the lake was very low. This lone reedbuck had wandered out onto an exposed sandbank, making for an excellent shot in the evening light.

Crossing a Shallow Lake **Bernard Castelein** 1999

I didn't choose my longest lens for this picture, but by putting this jackal rather small in the frame, I set out to suggest its easy walk almost 'on' the surface of the water.

Sunset at Mana Pools **Fanie Kloppers** 1996

One evening we took a drive on one of the very sandy roads at Mana Pools in Zimbabwe, and came across a big herd of buffalo. The buffalo were walking towards us and I wanted to see where they were heading, so I looked over my shoulder and saw the sun shining through the trees. I then noticed an impala walking towards the middle of these trees and knew that if it would stand still it would make an incredible picture. Luckily it happened exactly like that.

Too High to Jump! **RJ van Vuuren** 1995

These frogs, of the genus *Hyperolius*, are very photogenic indeed. The photograph was taken at St Lucia in KwaZulu-Natal on a hot humid night, using a single off-camera flash.

Cannibalistic Bullfrog **Nico Smit** 1998

After metamorphosis, young bullfrogs leave the water in large numbers in search of food to support them through the winter. Severe competition for food often results in cannibalism. When I came upon this young bullfrog at Warmbaths in South Africa's Northern Province, I was amazed at its ability to swallow a red toad almost its own size.

Two's Company **Johann Visser** 2000

While on a hike in the Tsitsikamma National Park, in South Africa's Eastern Cape province, I noticed these two grasshoppers in a carpet of pink flowers. Focussing was made difficult by a strong wind pushing and pulling the grasshoppers out of focus, so I held my breath and took a photo every time the wind calmed down.

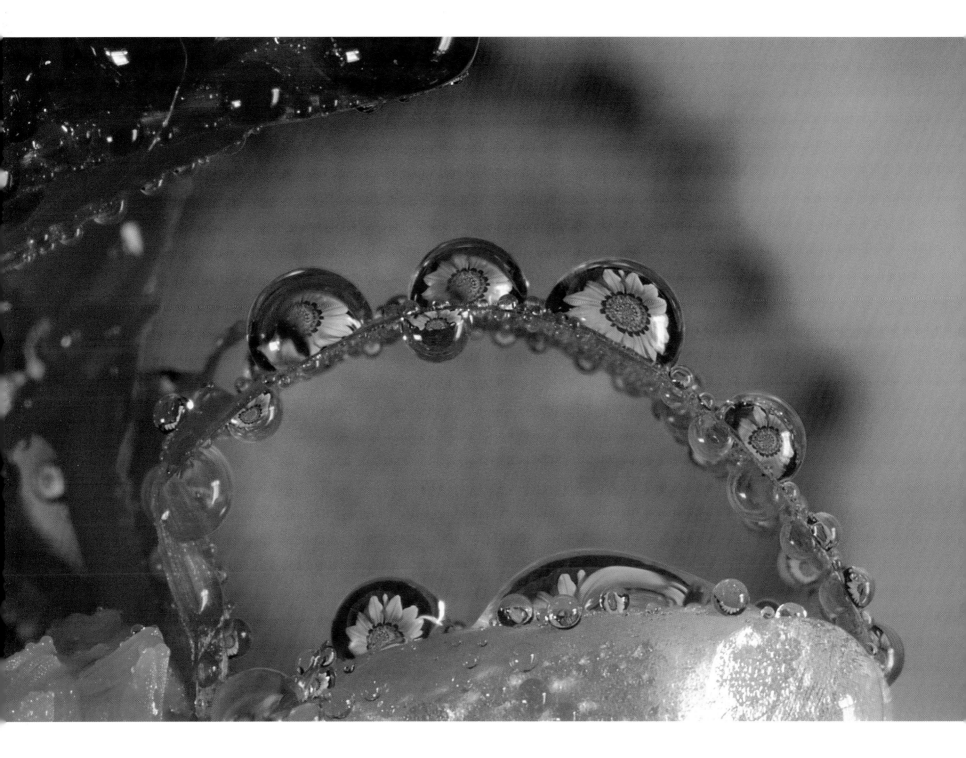

Dewdrop Image **JL Du Plessis** 2000

My interest in macro photography led me to droplet images.
Fascinated by this subject, I photographed a series of indigenous
flower images in dewdrops, on spiderwebs and on flower stamens.
Once I mastered the technicalities, it was like opening a jewelbox of
colours, and I was transported into this magical world of tiny baubles.

Tree Meal **James Wakelin** 2000

We were following a pride of lions in the Sabi Sand Game Reserve, when the pride came across a warthog carcass in a marula tree, presumably left there by a leopard. In a most undignified and comical manner, the entire pride scaled the tree and proceeded to feast upon the carcass.

Leopard with Kill Climbing a Tree
Duncan Usher 2000

We discovered this leopard resting in a shady tree after a successful hunt. Its kill (an impala) had fallen onto the ground. Later in the afternoon he attempted to drag the impala up another tree. After the third attempt, he was successful.

144

Photographers' Details and Technical Information

ALLOFS, THEO
Allofs Photography, P O Box 5473, Haines Junction, Yukon, Canada Y0B 1L0; Fax: +1 (867) 634 2207; Email: allofsphoto@yknet.yk.ca; *Black-backed Jackal Behind Elephant Leg* 1998 (pp 40/41); CAMERA: Nikon F5; LENS: 600 mm (f4); *Elephant Taking Mud Bath* 1998 (pp 96/97); CAMERA: Nikon F5; LENS: 600 mm (f4); *Moon between Rocks, Spitzkoppe, 2000* (p 24); CAMERA: Nikon F4S; LENS: 600 mm; *Namaqualand Wild Flowers* 1998 (p 70); CAMERA: Nikon F5; LENS: 105 mm (f2.8) macro.

ANDERS, UWE
Rosenstrasse 10, 38102, Braunschweig, Germany; Tel/Fax: +49 (531) 71374; Copyright © Uwe Anders/Team Dr Lammers; Dr R Lammers: Ketterler Strasse 5, 33415, Verl, Germany; Fax: +49 (524) 63166; *Black Rhino* 1993 (p 46); CAMERA: Nikon F4; LENS: 300 mm; EXP.: f4 at 1/125 sec.

ANNICIELLO, C
P O Box 782297, Sandton, 2146, South Africa; *This Little Piggy...* 1993 (p 31); CAMERA: Canon; LENS: 300 mm; EXP.: f4 at 1/125 sec.

BAILEY, ADRIAN
P O Box 1846, Houghton, 2041, South Africa; Fax: +27 (11) 805 8147; *Wildebeest* 1998 (p 85); CAMERA: Nikon F90X; LENS: 500 mm.

BARR, WILLIAM
5 Fairleads, Danbury, Essex CM3 4PR, UK; Email: bill-barr@wildlife-photography.co.uk; *Egyptian Goose* 1998 (p 23); CAMERA: Canon EOS; LENS: 300 mm.

BEYERS, JOHAN
5 Kleinmeul Street, Paarl, 7646, South Africa; Fax: +27 (21) 862 2192; Email: jbeyers@iafrica.com; *Great Egret* 1996 (p 134); *Heron Bathing* 1996 (p 23); CAMERA: Canon EOS 1N; LENS: 600 mm (f4); EXP.: f4 at 1/1000 sec.

BRAZENDALE, JOHN
Email: docks@iafrica.com; *Rush Hour Traffic* 1993 (p 12).

BROOKS, JJ
Chilanga, Parel Valley Road, Somerset West, 7130, South Africa; *Owl Landing* 1990 (p 11).

BRUWER, MARIKE
P O Box 255, Piketberg, 7320, South Africa; Fax: +27 (22) 913 2597; Email: ferdburg@intekom.co.za; *Namib Desert* 1997 (p 58); CAMERA: Nikon F801S; LENS: 24 mm, polariser; EXP.: f16.

BURGER, MARIUS
27 Risi Street, Fish Hoek, 7975, South Africa; *Kermit was Here* 1997 (p 77); CAMERA: Minolta X300; LENS: 50 mm with 2 x macro converter combination; FLASH: National PE-145; EXP.: f11 at 1/60 sec.

CALORE, MARCO
12 Edward Drive, Glendower, Edenvale, 1609, South Africa; *Long-crested Eagle* 1998 (p 94); CAMERA: Pentax Z1P; LENS: 80-200 mm (f2.8); FLASH: fill in flash; EXP.: f5.6 at 1/30 sec.

CAREW, DAPHNE
P O Box 206, Germiston, 1400, South Africa; Fax: +27 (11) 828 9908; *Skirmish* 1989 (p 11); CAMERA: Nikon EM; LENS: 600 mm; EXP.: f5.6 at 1/500 sec.

CAREW, TERRY
P O Box 206, Germiston, 1400, South Africa; Fax: +27 (11) 828 9908; *Forced Landing* 1986 (p 10); CAMERA: Nikon F3;

LENS: 600 mm; EXP.: f8 at 1/500 sec.; *Red Lechwe and Young* 1994 (p 45); CAMERA: Nikon F3; LENS: 600 mm; EXP.: f8 at 1/125 sec.; *Robber Fly* 1985 (p 78); CAMERA: Nikon F3; LENS: Nikkor Micro 105 mm; FLASH: Metz 60 CT-1; *Ten Little Bee-eaters in a Row* 1994 (p 34); CAMERA: Nikon FM2; LENS: Nikkor IFED 400 mm; FLASH: Metz 60 CT-1.

CASTELEIN, BERNARD
Verhoevenlei 100, B-2930 Brasschaat, Belgium; Tel/Fax: +32 (3) 653 08 82; *Common Agama* 1999 (p 49); CAMERA: Nikon F4; LENS: 200 mm macro; EXP.: f11; *Crossing a Shallow Lake* 1999 (p 137); CAMERA: Nikon F4; LENS: 300 mm; EXP.: f5.6.

DALTON, JOHN
St Lucia Reedbuck 1993 (p 136).

DE KOCK, TREVOR
Pelican Pictures S A c.c., Willow Valley, 5 Tamarisk Lane, Constantia, 7806, South Africa; Fax: +27 (21) 794 1876; Email: pelpic@iafrica.com; *Karoo Bigfoot* 1993 (p 93).

DE LA HARPE, PAT
P O Box 912, Howick, 3290, South Africa; Tel/Fax: +27 (33) 330 2789; Email: roger@africaimagery.co.za; *River Croc and Hatchling* 1996 (pp 26/27); CAMERA: Canon EOS 100; LENS: 300 mm (f4) IS with 1.4 x converter.

DE LA HARPE, ROGER
P O Box 912, Howick, 3290, South Africa; Tel/Fax: +27 (33) 330 2789; Email: roger@africaimagery.co.za; *Keeping Watch (Only Just!)* 1996 (p 119); CAMERA: Canon EOS 5; LENS: 300 mm (f2.8) with 1.4 x converter; EXP.: 1/15 sec.; *Sooty Tern Colony* 1999 (p 13); CAMERA: Canon EOS 1N; LENS: 17-35 mm; FLASH: 540 EZ; EXP.: 1/8 sec.; *Web-footed Gecko* 1999 (p 56); CAMERA: Canon EOS 1N; LENS: 100 mm macro.

DELPORT, KOOS
P O Box 38360, Faerie Glen, 0043, South Africa; *African Wildcat* 1989 (p 94); CAMERA: Nikon FE; LENS: 600 mm; EXP.: f5.6 at 1/500 sec.; *Airborne Impala* 1981 (pp 50/51); CAMERA: Nikon FE; LENS: 400 mm; EXP.: f5.6 at 1/500 sec.; *Cheetah Charge* 1985 (winner) (p 10); CAMERA: Nikon FE; LENS: 500 mm; EXP.: f11 at 1/125 sec.

DENIS-HUOT, M & C
1 bis rue des Fermes, 76310, Sainte-Adresse, France; Fax: +33 (2) 35 480430; Email: denishuot@aol.com; *Leopard Yawning* 1999 (p 16); CAMERA: Canon EOS 1N; LENS: 300 mm (f2.8) EF; *Lion Carrying Cub* 1999 (p 26); CAMERA: Canon EOS 1N-RS; LENS: 600 mm (f4) EF.

DENNIS, NIGEL J
Wildlife Photography, P O Box 580, Howick, 3290, South Africa; Tel/Fax: +27 (33) 330 7238; Email: ndwp@mweb.co.za; *Lion Yawning* 1998 (p 95); CAMERA: Canon EOS 1N; LENS: 600 mm (f4); EXP.: f4 at 1/125 sec.; *Namaqua Chameleon* 1998 (p 107); CAMERA: Canon EOS 1N; LENS: 20-35 mm zoom with polariser; FLASH: fill-flash; EXP.: f16 at 1/15 sec.; *Springbok Herd* 1998 (p 116); CAMERA: Canon EOS 1N; LENS: 600 mm (f4); EXP.: f4 at 1/15 sec.

DE VILLIERS, ELANIE
P O Box 403, Rustenburg, 0300, South Africa; Email: andreii@mweb.co.za; *Alert Cheetah* 1998 (p 107); CAMERA: Canon; LENS: 500 mm (f4.5) with 2 x extender.

DONALDSON, PAT
P O Box 662, Hoedspruit, 1380, South Africa; Tel/Fax: +27 (015) 793 2212; *Fatal Attraction* 1992 (p 37); CAMERA: Nikon F90X; LENS: 80-200 mm(f2.8).

DREYER, GERHARD
c/o Eugenie Walter, P O Box 25, Yzerfontein, 7351, South Africa; Tel/Fax: +27 (22) 451 2650; Email: emmaus@fast.co.za; *Colour and Cloud* 1994 (p 104); CAMERA: Nikon F801; LENS: 28-70 mm.

DU PLESSIS, GPL
Glenthorpe Plantation, P O Box 298, Barberton, 1300, South Africa; Fax: +27 (13) 712 3782; Email (c/o): janices@za.sappi.com; *Namib Kuns* 1982 (p 21); CAMERA: Nikon F3; LENS: 75-300 mm zoom; EXP.: f22 at 1/15 sec.

DU PLESSIS, JL
21 Buitekant Street, Swellendam, 6740, South Africa; *Dew Drop Lenses* 1999 (imprint and contents pages); CAMERA: Nikon F4; LENS: reversed 50 mm with bellow extension; FLASH: 2 flash heads; *Dewdrop Image* 2000 (p 143); CAMERA: Nikon F4; LENS: 50 mm micro with bellow extension; FLASH: SB16 and SB23 flash units.

DU TOIT, RICHARD
P O Box 547, Witkoppen, 2068, South Africa; Tel/Fax: +27 (11) 465 9919; Email: rdutoit@iafrica.com; *Elephant Shrew* 1997 (p 76); CAMERA: Canon EOS 1N; LENS: 500 mm; EXP.: f6.3 at 1/640 sec.; *Rage* 1999 (p 36); CAMERA: Canon EOS 1N; LENS: 300 mm; EXP.: f4 at 1/125 sec.

FERREIRA, HENDRIK
P O Box 984, Melville, 2109, South Africa; Tel/Fax: +27 (11) 792 3956; Email: hcf@ing1.rau.ac.za; *Interleaved Dead Trees* 1999 (p 20); CAMERA: Minolta X700; LENS: 70-300 mm zoom; EXP.: f16.

FRANKLIN, PETER
11 Lindeshof Road, Constantia Hills, Cape Town, 7806, South Africa; Fax: +27 (21) 425 2369; Email: franklindi@yebo.co.za; *Silent Retreat* 1987 (p 10).

FRONEMAN, ALBERT
P O Box 242, Cramerview, 2060, South Africa; Tel (w): +27 (11) 486 1102; *Spur-winged Surfer* 1997 (p 22); CAMERA: Nikon F90; LENS: 500 mm with 1.4 x extender; EXP.: f5.6 at 1/125 sec.

GOSS, RICHARD
P O Box 2562, White River, 1240, South Africa; Fax: +27 (13) 751 3023; Cell: +27 (82) 556 0287; *Desert King* 1983 (Ltd ed. prints available) (p 9); CAMERA: Olympus; LENS: 50 mm.

GRABEN, MATTHIAS
Everstalstr. 86, D-44894 Bochum, Germany; Tel: +49 (0) 234 264 174; *Shadows on Namib Dunes* 1998 (p 59); CAMERA: Nikon F4S; LENS: 80-200 mm; EXP.: f8 at 1/60 sec.

GREAVES, NICK
P O Box 54, Esigodini, Zimbabwe; *Crocodile Attack* 1996 (p 12); CAMERA: EOS 1; LENS: 300 mm (f4).

GRIFFITHS, WAYNE
P O Box 13602, Northmead, 1511, South Africa; Fax: +27 (11) 845 4585; Email: wayne@fleetsure.co.za; *Aggression* 1995 (p 69); CAMERA: Nikon F4; LENS: 80-200 mm (f2.8); EXP.: f4 at 1/125 sec.

HAACKE, WD
Transvaal Museum, P O Box 413, Pretoria, 0001, South Africa; Email: haacke@nfi.co.za; *Survivor* 1984 (p 9); CAMERA: Ashahi Pentax Spotmatic; LENS: SMC Takumar 45-125 mm zoom; EXP.: f16 at 1/60 sec.

HARVEY, MARTIN
P O Box 8945, Centurion, 0046, South Africa; Tel/Fax: +27 (12) 664 2241; Email: mharvey@icon.co.za; *Johnstone's Chameleon* 1999 (p 77); CAMERA: Canon EOS A2; LENS: 100 mm macro; *Leopard on Termite Mound* 1999 (p 55); CAMERA: Canon EOS 1N; LENS: 300 mm; *Leopard with Baboon Kill* 1999 (p 98); CAMERA: Canon EOS 3; LENS: 70–200 mm; *Parson's Chameleon Eye* 1998 (pp 64/65); CAMERA: Canon EOS A2; LENS: 100 mm macro; FLASH: two flashes.

HELDRING-HAMMAN, HELENE
Private Bag 159, Maun, Botswana; Tel/Fax: +267 661 230; *Fact of Life* 1995 (p 68); CAMERA: Canon T-90; LENS: 28 mm; EXP.: f5.6 at 1/250 sec.

HES, LEX
P O Box 19113, Nelspruit, 1200, S A; Email: lexhes@global.co.za; *The King* 1988 (p 10).

HEY, MALCOLM
54 The Thicket, Romsey, Hampshire, SO51 5SZ, England, UK; Tel/Fax: +44 (1794) 513248; Email: mdh@malcolmhey.co.uk; *Blenny in its Coral Home* 1999 (p 115); CAMERA: Nikon F90-X; LENS: 105 mm macro; FLASH: dual Sea & Sea YS-50 flashguns; EXP.: f22.

HEYMANS, PIET
P O Box 6764, Bloemfontein, 9300, South Africa; Fax: +27 (51) 432 6545/95; Email: pieth@pixie.co.za; *Mother Care* 1993 (p 45); *Ousus* 1994 (p 63); CAMERA: Minolta 7xi; LENS: 600 mm (f4) with 1.4 x converter; EXP.: f5.6.

HILL, RAY
104 Parklands, Jarvis Road, Berea, East London, 5241, South Africa; *Knysna Lourie* 1983 (p 100); CAMERA: Minolta XG 2.

HINDE, GERALD
P O Box 9158, Minnebron, 1549, South Africa; Email: hinde@netactive.co.za; *Lioness Stalking* 2000 (winner) (p 6); CAMERA: Canon EOS 1N-RS; LENS: 600 mm; EXP.: f8 at 1/320 sec.

HIRSCHOWITZ, SETH
Level 3, 430 Victoria Avenue, Chatswood NSW 2067, Sydney, Australia; Fax: +61 (2) 913 05806; *The Duel* 1996 (p 126); CAMERA: Nikon N90; LENS: Tokina 150–500 mm; EXP.: f8 at 1/60 sec.

HOPKINS, GARETH
Cell: +27 (83) 481 3675; Email: driehoek@dullstroom.net; *New Beginnings* 2000 (p 113); CAMERA: Nikon F90X; LENS: 105 mm (f2.8) micro; *Rhinos at Sunset* 2000 (p 129); CAMERA: Leica R7; LENS: Leitz 280 mm (f2.8).

HUNTER, LUKE
ABC Natural History Unit, 10 Selwyn Street, Elsternwick, UIC 3185, Australia; Email: hunterluke@hotmail.com; *Cheetah Cub Lookout* 1995 (p 54); CAMERA: Canon EOS 5; LENS: 80–200 mm zoom; EXP.: f2.8 at 1/125 sec.

JACKSON, TIM
Department of Zoology & Entomology, University of Pretoria, Pretoria, 0002, South Africa; Email: tjackson@zoology.up.ac.za; *Flycatchers in the Cold* 1994 (p 35); CAMERA: Nikon F801S; LENS: 400 mm with 1.4 x teleconverter; EXP.: f8 at 1/250 sec.

JACOVIDES, MARIA
Email: ttsonis@mweb.co.za; *High Speed Chase* 1995 (p 12); CAMERA: Canon EOS 650; LENS: Sigma 75–300 mm.

JESCHE, FRANZ
P O Box 1848, Randburg, 2125, South Africa; Fax (w): +27 (11) 787 4080; *Sunset Vigilance* 1997 (p 133); CAMERA: Leica R7; LENS: 400 mm; EXP.: f2.8 at 1/250 sec.

JOOSTE, JOHAN
P O Box 9661, Windhoek, Namibia; Fax: +264 (61) 243 219; Email: jjooste@mweb.com.na; *Annoyed Elephant* 1992 (p 16); CAMERA: Minolta 9xi; LENS: 80–200 mm zoom with polariser; EXP: f8 at 1/250 sec.

JOUBERT, BEVERLY
c/o National Geographic Image Sales; Email: image@ngs.org; *Giant Kingfisher Catching a Fish* 1985 (p 22); CAMERA: Nikon F4; LENS: 400 mm; EXP: f5.6; *Out of Control* 1995 (p 14);

CAMERA: Canon EOS; LENS: 300 mm; EXP.: f5.6; *Quelea Dance* 1994 (pp 72/73); CAMERA: Nikon F4; LENS: 500 mm; EXP: f5.6 at 1/250 sec.; *Surprise Encounter* 1995 (p 31); CAMERA: Canon EOS; LENS: 600 mm (f8).

KITZEN, MARTIN
245 St Andrews Road, Epsom, Auckland, 1003, New Zealand; Email: martin@sanzglobal.co.nz; *Heaven Scent* 1995 (p 132); CAMERA: Canon F1; LENS: Tokina 150–500 ATX zoom.

KLOPPERS, FANIE
Leopard Photo Enterprises; Tel: +27 (83) 270 7269; *Sunset at Mana Pools* 1996 (pp 138/139); CAMERA: Nikon F3; LENS: 100–300 mm zoom.

KLOPPERS, JOHAN
Leopard Photo Enterprises; Tel: +27 (83) 270 7269; *Leopard* 1983 (front cover; pp 122/123); CAMERA: Leica; LENS: 560 mm telephoto.

KNOBEL, JOHANN
P O Box 15775, Sinoville, 0129, South Africa; Email: knobecj@unisa.ac.za; *Stampede* 1992 (full title); CAMERA: Nikon FM2; LENS: Novolflex 600 mm; EXP.: f8 at 1/500 sec.

KRAHMER, FRANK
Bahnhofstr. 13 F, Taufkirchen, Germany; Fax: +49 (89) 61209243; Email: fkrahmer@compuserve.com; *African Buffalo* 1999 (pp 18/19); CAMERA: Canon EOS 1N; LENS: 500 mm (f4.5) EF with 1.4 x converter; EXP.: f4.5 at 1/250 sec.

LAMBERTI, STEFANIA
P O Box 11, Buccleuch, 2066, South Africa; Email: stefania@icon.co.z; *Plunging into the Deep Blue* 1999 (pp 90/91); CAMERA: Nikon 801; LENS: 20 mm in Sea Cam housing; EXP.: f8 at 1/125 sec.

LAMPRECHT, GERT
P O Box 28069; Danhof, Bloemfontein, 9310, South Africa; Email: lampregj@cem.nw.uovs.ac.za; *Mtunzini Frog* 1995 (p 15); CAMERA: Canon EOS 5; LENS: 100 mm macro; EXP.: f22 at 1/200 sec.

LIGHTBODY, EDWARD
P O Box 650019, Benmore, 2010, South Africa; *After the Mating* 1981 (p 9); CAMERA: Pentax; EXP.: f8 at 1/250 sec.

LILLIE, PETER
C 204 Devonshire Hill, 13 Grotto Road, Rondebosch, 7700, Cape Town, South Africa; *Suricate with Young* 1992 (p 60); CAMERA: Nikon F3; LENS: 600 mm; EXP.: f8 at 1/250 sec.

LIVINGSTONE, MJ
P O Box 182, Simon's Town, 7995, Western Cape, South Africa; Email: livingst@mweb.co.za; *Ibis Feeding on Maggots* 1999 (p 125); CAMERA: Nikon F5; LENS: 400 mm (f2.8) with 1.4 x converter; EXP.: f8 at 1/250 sec.; *Terns* 2000 (p 112); CAMERA: Nikon F5; LENS: 400 mm (f2.8); EXP.: f16 at 1/125 sec.

LOOTS, HENK
P O Box 92287, Mooikloof, 0059, South Africa; Tel/Fax: +27 (012) 996 0099; *A Tree in the Desert* 1995 (p 57); CAMERA: Minolta X-700; LENS: Sigma 75–300 mm APO; EXP.: f8.

LOUW, JOHAN
208 Kalipiestraat, Pretoria Noord, 0182, South Africa; *Eekhoring* 1996 (p 76); CAMERA: Nikon F90X; LENS: Tokina 150–500 zoom with 1.4 x converter; EXP.: f11 at 1/15 sec.

MALAN, CORNIE
P O Box 236, Dundee, 3000, South Africa; *Hokaai!* 1996 (p 39); CAMERA: Canon T90; LENS: Tokina 100–300 mm; EXP.: 1/30 sec.

MATTHEWS, WAYNE
Tembe Elephant Park, Private Bag, X356, KwaNgwanase, South Africa; Fax: +27 (35) 592 0240; *Last Light Drink* 2000 (p 40); CAMERA: Canon EOS 1N; LENS: 100–400 mm IS.

MEAD, COLIN
P O Box 68914, Bryanston, 2021, South Africa; Tel/Fax: +27 (11) 706 5360; Email: margmead@icon.co.za; *After the Floods* 1997 (p 102); CAMERA: Canon EOS 1N; LENS: 28–105 mm zoom with polariser; EXP.: f22. at 1/8 sec.

MICHLER, IAN
Private Bag 23, Maun, Botswana; Fax: +267 678 015; *Black Egret Shading* 1998 (p 135); CAMERA: Nikon N90S; LENS: 300 mm; EXP.: f5.6 at 1/125 sec.

MIGDOLL, IVOR
11 River Drive, Carrington Heights, Durban, 4001, South Africa; *Cheetah Chasing Springbok* 1996 (p 117); CAMERA: Pentax PZ1; LENS: 250–600 mm; EXP.: f5.6 at 1/60 sec.; *Goliath Heron* 1993 (p 135); CAMERA: Pentax PZ1; LENS: 250–600 mm; EXP.: f8 at 1/250 sec.

MILLER, LEN
10-11th Avenue, Orange Grove, 2192, Gauteng, South Africa; *Sandstorm Fantasy* 1992 (p 86); CAMERA: Nikon F3; LENS: 200 mm; EXP.: 1/125 sec.

MILLS, GUS
Private Bag X402, Skukuza, 1350, South Africa; Email: gusm@parks-sa.co.za; *A Bloody Mess* 1983 (p 69); CAMERA: Minolta X-GM; LENS: 200 mm; FLASH: Braun f800; *Locked Horns* 1982 (p 9); CAMERA: Minolta X-GM; LENS: Rokkor 50 mm.

MOSTERT, COLIN
c/o 8 Humby Road, Ottery, 7700, Cape Town, South Africa; Fax: +966 (3) 840 5183; Email: colinmostert@prime.net.sa; *Ambush* 1998 (p 32, bottom); CAMERA: Nikonos V; LENS: 35 mm with 1:1 extension tubes; FLASH: SB103 strobe; EXP.: f22 at 1/90 sec.; *Big Eyes* 1998 (p 33, top); CAMERA: Nikonos V; LENS: 35 mm with 1:1 extension tubes; FLASH: SB103 strobe; EXP.: f22 at 1/90 sec.; *Coral Kingdom* 1995 (p 105); CAMERA: Nikonos V; LENS: 15 mm; FLASH: SB103 strobe; EXP.: f22 at 1/90 sec.

MYBURGH, NICO
17 Jacobus Gildenham Street, Onrus River, 7201, South Africa; *Antics* 1994 (p 48); CAMERA: Novoflex; LENS: 400 mm; FLASH: 2 Metz flash heads.

NIEBUHR, HELMUT
P O Box 9223, Edleen, 1625, South Africa; Fax: +27 (11) 391 5401; *A Leisurely Lookout* 1996 (p 62); CAMERA: Canon EOS 1N; LENS: 600 mm (f4) with 1.4 x converter; *Cheetah* 1995 (p 38); CAMERA: Canon EOS 1N; LENS: 500 mm (f4.5); *Jackal Feud* 1996 (p 37); CAMERA: Canon EOS 1N; LENS: 600 mm (f4) with 1.4 x converter; *Tawny-flanked Prinia* 1992 (p 100); CAMERA: Nikon 801; LENS: Tamron 400 mm (f4) with 1.4 x converter.

NUNNINGTON, ROB
959 C Great View Circle, Dayton, Ohio 45459, USA; Email: rnunnington@yahoo.com; *Gemsbok and Calves* 1991 (p 11). CAMERA: Nikon F3HP LENS: 400mm f3.5 IFED EXP: f4 at 1/250th sec.

OATES, LES
P O Box 90310, Garsfontein, 0042, South Africa; *Salticidae in Lair* 1994; CAMERA: Pentax ES; LENS: 50 mm macro with extension tubes; FLASH: Braun lens.

PICKFORD, BEVERLY
P O Box 3, Hammanshof, 6863, South Africa; Fax: +27 (23) 342 1992; *Lion Charge* 1998 (p 13); *Red Lechwe Crossing Flood Plain* 1997 (back cover; p 75); CAMERA: Nikon F4; LENS: 80–200 mm zoom; EXP.: f5.6 at 1/500 sec.

PICKFORD, PETER
P O Box 3, Hammanshof, 6863, South Africa; Fax: +27 (23) 342 1992; *Dog Kill* 1997 (pp 80/81); CAMERA: Nikon F4; LENS: 600 mm; EXP.: f4 at 1/250 sec.; *Sunset with Grass* 1999 (half title page); CAMERA: Nikon F4; LENS: 105 mm macro; EXP.: f4 at 1/500 sec.

PINNOCK, PETER
P O Box 70138, Overport, Durban, 4067, South Africa; Fax: +27 (31) 204 8669; Email: peter.pinnock@spl.co.za; *Open Wide* 1998 (p 115); CAMERA: Nikon F90; LENS: 105 mm inside an Ikelite housing; EXP.: f8 at 1/60 sec.

PLAKONOURIS, IKY
P O Box 15574, Emerald Hill, Port Elizabeth, 6011, South Africa; Email: bushbabi@icon.co.za; *The King* 1992 (p 124); CAMERA: Nikon F3; LENS: 400 mm (f3.5) with 1.4 x converter; EXP.: f5.6 at 1/250 sec.

RANKIN, AJ
Private Bag 14, Maun, Botswana; Email: alrankin@info.bw; *Buffalo Sunset* 1996 (p 128); CAMERA: Canon EOS 1; LENS: 300 mm; EXP.: f5.6 at 1/250 sec.

ROONEY, MICHAEL J
Rooney's Hire Service, P O Box 1351, Harare, Zimbabwe; Email: zimhire@cst.co.zw; *Waterbuck Challenge* 1995 (p 127); CAMERA: Nikon F4; LENS: 28–200 mm zoom; EXP.: f2.8.

RYDER, JOAN
P O Box 2540, Bedfordview, 2008, South Africa; Fax: +27 (11) 824 4829; Email: jryder@yebo.co.za; *Hamerkop Scratching* 1985 (p 79); CAMERA: Canon T90; LENS: 150–600 mm (f5.6); *Migratory Butterfly* 1985 (p 120); CAMERA: Nikon F2; LENS: 100 mm macro; FLASH: flashgun; EXP.: f16.

SNEESBY, JILL
J & B Photographers, P O Box 5060, Walmer, 6065, South Africa; Fax: +27 (41) 581 1217; Email: bwilkins@iafrica.com; *African Honey Bee* 1995 (p 67); CAMERA: Canon EOS 1; LENS: 100 mm (f2) macro; FLASH: ring flash; *Baby Elephant* 1996 (p 118); CAMERA: Canon EOS 1; LENS: 600 mm (f4) with 1.4 x converter; *Elephant and Gemsbok* 1998 (p 47); CAMERA: Minolta X700; LENS: 100–300 mm; *Gemsbok Against Dunes* 1987 (p 106); CAMERA: Minolta X700; LENS: 100–500 mm; *Homeward Bound* 1985 (p 87); CAMERA: Minolta X700; LENS: 100–500 mm (f8); *Lioness Wading* 1997 (p 74); CAMERA: Canon EOS 1; LENS: 600 mm (f4); *Springbok Death* 1994 (p 127); CAMERA: Canon EOS 1; LENS: 600 mm (f8).

SPIBY, GEOFF
'Springtide', Lentedal Road, Hout Bay, 7806, South Africa; Fax: +27 (21)790 5946; *Goldies* 1999 (p 91, right); CAMERA: Nikonos IV-A; LENS: 15 mm with Ikelite 300 underwater strobe on 1/4 power; EXP.: f8 at 1/60 sec.; *Hunter and the Hunted* 1999 (p 90, left); CAMERA: Nikonos V; LENS: 15 mm; EXP.: f5.6–f5.8 at 1/60 sec.

STAEBLER, GABRIELA
Am Eichet 2, 86938, Schondorf, Germany; Fax: +49 8192 1095; *Cheetah Cub on Tree* 1998 (p 131); CAMERA: Canon EOS 1; LENS: 300 mm; EXP.: f5.6 at 1/60 sec.; *Cheetah Mother with Cub* 1996 (p 60); CAMERA: Canon T90; LENS: 500 mm; EXP.: f4.5 at 1/90 sec.; *Jackal Defending Skeleton* 1998 (p 110); CAMERA: Canon EOS 1; LENS: 300 mm; EXP.: f2.8 at 1/250 sec.; *Two Lion Brothers* 1996 (pp 52/53); CAMERA: Canon EOS 1; LENS: 600 mm; EXP.: f4.5 at 1/500 sec.

STECKBAUER, SCOTT S
625 Woodland Avenue, Chula Vista, CA 91910, USA; *Pounce* 1992 (p 130); CAMERA: Nikon FE-2; LENS: 400 mm (f3.5) with 1.4 x converter; EXP.: f3.5 at 1/125 sec.

TARBOTON, WARWICK
P O Box 327, Nylstroom, 0510, South Africa; Tel/Fax: +27 (14) 743 1438/1442; *Masked Weaver* 1994 (p 101); CAMERA: Canon EOS Elan; LENS: 300 mm (f2.8).

THOM, JAMIE
P O Box 763, Strathavon, 2031, Sandton, South Africa; *Sad, Wet and Sorry* 1994 (p 61); CAMERA: Nikon F4; LENS: 80–200 mm (f2.8); EXP.: f2.8 at 1/60 sec.; *Feeding Frenzy* 1994 (pp 108/109); CAMERA: Nikon 801S; LENS: 75–300 mm; FLASH: Metz 60; EXP.: f5.6 at 1/125 sec.

TOON, ANNE & STEVE
6 Lynslack Terrace, Arnside, Cumbria LA5 0EL, UK; Tel/fax: +44 (1524) 762 804; Email: sandatoon@aol.com; *Adult Hyaena with Pup* 1998 (p 27); CAMERA: Canon EOS 1N; LENS: 300 mm; EXP.: f5.6 at 1/125 sec.

URQUHART, COLIN
11 Bluecliff Avenue, Bridgemeade, Port Elizabeth, 6025, South Africa; Email: cbu@intekom.co.za; *Flower Power* 1995 (p 66); CAMERA: Canon A1; LENS: Vivitar 100–500 mm (f5.6); EXP.: f8 at 1/250 sec.

USHER, DUNCAN
Muehlenbreite 9, 34346 Hann. Munden (Bursfelde), Germany; Fax: +49 (0) 5544 912020; *Leopard with Kill Climbing a Tree* 2000 (p 145); CAMERA: Canon EOS 3; LENS: 600 mm (f4) with 1.4 x converter.

VAN DEN BERG, HEINRICH
P O Box 13244, Cascades, 3202, South Africa; *Springbok in Rain* 1997 (pp 28/29); CAMERA: Canon EOS 5; LENS: (f2.8).

VAN DEN BERG, HERMAN
P O Box 13244, Cascades, 3202, South Africa; *Black Wildebeest and Newly Born Calf* 1994 (p 44); CAMERA: Canon T90; LENS: 500 mm (f4.5); *Dandelion Precision* 1995 (endpages); CAMERA: Canon T90; LENS: 100 mm macro; EXP.: f16; *Lion Cub* 1997 (p 130); CAMERA: Canon EOS 5; LENS: 300 mm (f2.8) with 2 x converter; EXP.: f5.6; *Pink-backed Pelican* 1994 (p 82); CAMERA: Canon T90; LENS: 500 mm (f4.5); *Red Hartebeest in Flight* 1994 (p 50); CAMERA: Canon T90; LENS: 500 mm (f4.5).

VAN DEN BERG, INGRID
P O Box 13244, Cascades, 3202, South Africa; *Fighting Giraffe Bulls* 1998 (p 92); CAMERA: Canon EOS 5; LENS: 300 mm (f2.8) with 1.4 x converter; EXP.: f4 at 1/125 sec.; *Newborn Baboon* 1994 (p 131); CAMERA: Canon F1; LENS: 300 mm.

VAN DEN BERG, PHILIP
P O Box 13244, Cascades, 3202, South Africa; *Feeding Vervet Monkey* 1997 (p 99); CAMERA: Canon EOS 5; LENS: 300 mm (f2.8); EXP.: f2.8; *Fiscal Shrike Perched on Aloe* 1999 (p 101); CAMERA: Canon EOS 5; LENS: 300 mm (f2.8) with 1.4 x converter; EXP.: f4 at 1/125 sec.; *Malachite Kingfisher with Catch* 1997 (p 100); CAMERA: Canon T90; LENS: 500 mm (f4.5) with extension ring; *Preening Spoonbill* 1993 (spine; p 83); CAMERA: Canon F1; LENS: 300 mm; FLASH: fill in flash; EXP.: f8.

VAN DER MERWE, THYS
Mobile: +27 (83) 441 3108; Email: teknovis@cis.co.za; Web: http://home.mweb.co.za/te/teknovis; *Spitzkoppe Arch* 2000 (p 25); CAMERA: Canon EOS 5; LENS: 15 mm (f2.8); EXP.: f8 at 1/60 sec.

VAN EEDEN, FREEK
P O Box 38328, Faerie Glen, 0043, South Africa; *Wet Lion* 1998 (p 84); CAMERA: Canon EOS 1; LENS: 600 mm; EXP.: f5.6 at 1/250 sec.

VAN EEDEN, REIMUND
38 Lily Avenue, Murrayfield, Pretoria, 0184, South Africa; Email: ray@prodivers.com; *Coral Hind* 1997 (p 33, bottom); CAMERA: Nikon F90; LENS: 105 mm; EXP.: f11. Equipment inside a Sealux housing with a subtronic amphibian flash.

VAN VUUREN, DR RJ
P O Box 284, Graskop, 1270, South Africa; *Hippo Feud* 1994 (p 12); CAMERA: Canon EOS 1; LENS: ultrasonic 600 mm (f4); EXP.: f5.6.; *Too High to Jump!* 1995 (p 140); CAMERA: Olympus OM-10; LENS: 50 mm on a Komura Telemore 7-element 2 x converter; EXP.: f16; *Watch Out!* 1997 (p 56); CAMERA: Olympus OM4 Ti; LENS: 80 mm macro on extension tube; FLASH.: two flashes; EXP.: f22.

VAN WYK, JACO
Hoërskool Waterkloof, P O Box 25085, Monument Park, Pretoria, 0105, South Africa; *Albatross Chick* 1995 (p 83); CAMERA: Pentax K1000; LENS: 28–80 mm; EXP.: f11 at 1/125 sec.

VISSER, JOHANN
P O Box 22155, Extonweg, Bloemfontein, 9313, South Africa; Email: johannvisser@hotmail.com; *Two's Company* 2000 (p 142); CAMERA: Minolta 600si; LENS: Sigma 70–300 mm (f4–f5.6) zoom; EXP.: f5.6 at 1/90 sec.

WAGNER, PATRICK
Copyright © Patrick Wagner/Photo Access; Tel: +27 (21) 531 4341; *Nose-stripe Clownfish, Mozambique*, 1997 (p 71); CAMERA: Nikonos V; LENS: close-up lens attachment with a Nikonos SB-102 strobe; *White-winged Terns, Uganda*, 1997 (p 42); CAMERA: Nikon F4; LENS: 200–300 mm; FLASH: SP24.

WAKELIN, JAMES
KZN Nature Conservation Service, Scientific Services, P O Box 13053, Cascades, 3202, South Africa; *Tree Meal* 2000 (p 144); CAMERA: Nikon F90X; LENS: 80–200 mm (f2.8); FLASH: speedlight flash; EXP.: f2.8 at 1/60 sec.

WALLINGTON, GRAHAM
Email: graham@africam.co.za; *Coral Trout* 1993 (p 32, top); CAMERA: Nikon F3; LENS: 600 mm with 2 x converter; EXP: f8 at 1/125 sec.

WEINBERG, JACK
P O Box 3780, Johannesburg, 2000, South Africa; *King of the Dunes* 1987 (p 103); CAMERA: Nikon F3; LENS: 600 mm with 1.4 x converter; EXP.: f6.6 at 1/60 sec.; *Mother and Foal* 1986 (p 119); CAMERA: Nikon F3; LENS: 600 mm and warming filter; *Red Bishop Calling* 1985 (p 101); CAMERA: Nikon F3; LENS: 600 mm with 2 x converter; EXP.: f8 at 1/125 sec.

WILKINS, BARRIE
J & B Photographers, P O Box 5060, Walmer, 6065, South Africa; Fax: +27 (41) 581 1217; Email: bwilkins@iafrica.com; *Pronking Springbok* 1984 (p 51); CAMERA: Leica R3; LENS: Pentax 135–600 mm zoom; EXP.: f8 at 1/15 sec.

WILLIAMS, DC
P O Box 35201, Northway, 4065, South Africa; Email: williamd@durban.gov.za; *Pink-backed Pelican Duo* 1997 (p 43); CAMERA: Minolta 9xi; LENS: 300 mm (f8) with 1.4 x converter.

WILSON, ALAN
2 Minerva Court, Minerva Grove, Durban, 4001, South Africa; *Pelican Take-off* 1992 (p 11); CAMERA: EOS 1; LENS: 600 mm (f4)

WOLHUTER, KIM
P O Box 1550, White River, Mpumalanga, 1240, South Africa; Fax: +27 (13) 750 1486; Email: wildhoot@mweb.co.za; *Black-backed Jackal* 1998 (p 13); CAMERA: Canon EOS 1; LENS: 35–350 mm; EXP.: f5.6 at 1/150 sec. ; *Zebras in a Panic in a Pan* 1999 (pp 88/89); CAMERA: Canon EOS 1; LENS: 35–350 mm; EXP.: f5.6 at 1/125 sec.

WOODS, KEN
5 Kenmar Crescent, Claremont, 7700, South Africa; Email: kwoods@iafrica.com; *Butterfly Perch* 1997 (p 121); CAMERA: Nikon F90S; LENS: 200 mm (f4) macro.

WU, NORBERT
1065 Sinex Ave, Pacific Grove, CA 93950, USA; Tel +1 (831) 375-4448; Web: http://www.norbertwu.com; Email: norboffice@aol.com; *Sea Star, Seychelles* 2000 (p 112); CAMERA: Canon F1; LENS: 50 mm macro with flat port; FLASH: strobe at full power; EXP.: f11 at 1/60 sec. Equipment in Oceanic housing.

ZIETSMAN, PC
P O Box 266, Bloemfontein, 9300, South Africa; Fax: +27 (51) 421 5881; *Koggelmanderwyfie* 1995 (p 114); CAMERA: Asahi Pentax ME Super; LENS: 100 mm macro; EXP.: f11 at 1/125 sec.